A Challenge

In algebra we use equations to solve problems that are hard
Here are some problems which can be solved easily using algebra. They can also be solved without using algebra. Work on these problems any way you want. Make a record of how you solved them. Later we will solve problems like these using algebra.

"I'm thinking of a number. If you multiply it by 6 and then add 7, you will get 55. What is my number?"

One number is 5 more than another. Their sum is 53. What are the two numbers?

A shake at the Shake Shack costs 80¢. The bill for three burgers and a shake was $4.40. How much is a burger?

Kim is 15. Kim's father is 39. In how many years will Kim's father be exactly twice as old as Kim will be?

You can ride a bicycle 15 mph. Your sister rides 12 mph. You give her a one hour head start. In how many hours will you catch up if neither of you stops?

The Rag Bag is having a "20% Off" sale. A pair of pants is on sale for $16. How much was it to begin with?

Equations

Here is an equation in x: $$10 + x = 3$$

We can **solve** this equation by substituting different numbers for x. If we find a number that works, then we have found a **solution** to the equation. The only number that will work in this equation is ⁻7. We can show the solution by writing

$$x = {}^-7$$

Solve each equation by substituting different numbers until you find a solution.

$5 + y = 9$ $y = 4$	$z + 1 = 10$	$3 + w = 4$	$a + 15 = 20$
$10 + x = 17$	$4 + t = 4$	$^-6 + r = {}^-16$	$8 + x = 6$
$c + 6 = 5$	$^-5 + y = {}^-1$	$c + {}^-8 = {}^-8$	$17 + a = 23$
$^-4 + a = {}^-9$	$^-3 + m = {}^-13$	$x + {}^-2 = 5$	$4 + e = {}^-4$

Joe picked a number and added 7 to it. The answer was 11. What was the number?

Equation: $x + 7 = 11$

Solution: $x = 4$

Jo picked a number and added 3 to it. The answer was 12. What was the number?

Equation:

Solution:

Chip thought of a number. He added 9 to it. The answer came out to be 14. What was Chip's number?

Equation:

Solution:

Key to
Algebra ®

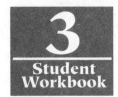

3
Student
Workbook

Equations

By Julie King and Peter Rasmussen

Name   Class

TABLE OF CONTENTS

A Challenge 1
Equations 2
Solving Equations 4
Equivalent Equations 6
The Finger Method 9
The Addition Principle for Equations 13
Using Equations to Solve Problems 24
Age Puzzles 27
Perimeter Problems 28
The Division Principle for Equations 29
The Multiplication Principle for Equations 32
Written Work 35
Practice Test 36

A Renaissance "Duel"

The Arabic scientist Mohammed ibn-Musa al-Khowarizmi wrote a book around 800 A.D. in which he systematically solved all quadratic equations. Quadratic equations are equations like $3x^2 + 7x - 6 = 0$ where the highest power of x is 2. Some scientists were able to extend his results. Omar Khayyam, who is known in the West as a poet but in the East as a great scientist, discovered a way to solve certain cubic equations. A cubic equation is an equation like $2x^3 + 5x^2 - x + 7 = 0$ where the highest power of x is 3.

Nobody was able to find a formula that would solve all cubic equations until the Renaissance. The story of that discovery is one of the most fascinating episodes in the history of mathematics.

The story begins with Scipione del Ferro, an Italian who discovered how to solve certain cubic equations. That was the biggest advance in algebra since Mohammed defined the subject some 700 years earlier.

But del Ferro didn't tell anyone about his discovery. Why not? Was he overly modest? No. On the contrary.

You see, in those days teaching positions were obtained by challenges. If someone felt smarter than the present teacher, then a challenge was issued. Each contestant had to solve problems posed by the other. The winner was the person who solved the most problems.

Del Ferro, a professor at Bologna University, kept his professorship by posing cubic equations to all challengers. Even if he couldn't solve a challenger's problems he knew that he could keep his job because nobody else could solve cubic equations. (The incumbent, not the challenger, won all ties.)

It is unlikely that you have ever heard of del Ferro. No wonder. He only told his formula to a few trusted friends, including his student Antonio Fior. Besides, some 19 years after his death a new book published a formula for all cubic equations, not just the ones del Ferro knew.

The story behind that book is fascinating. It begins with Niccolo Tartaglia, who heard the rumor of del Ferro's accomplishment and somehow learned the formula. He was rather offensive and boastful, however, and Fior challenged him to a duel.

Now, we hasten to inform you that such duels were not fought with guns or swords. They were contested in public, usually in a courtyard. But the weapons were mathematical problems.

In Italian the word *tartaglia* means *stutterer*, a nickname Niccolo got from a speech impediment incurred from a boyhood accident with a sword. He didn't need a sword to win this duel, however, because on the eve of it he discovered a way to solve cubic equations beyond the kind that del Ferro's method could handle.

Tartaglia's intention was to publish a book containing his secret formula in order to win the support of a wealthy patron, like the well known artists Leonardo and Michelangelo had. But he never wrote the book.

In the meantime he told his formula to a wealthy physician, Jerome Cardano, who, some five years later in 1545, wrote a book, *The Great Art*, containing the secret formula. Tartaglia was incensed because Cardano had promised not to reveal the formula to anyone. Cardano responded that he published the book because Tartaglia had never written his promised book and because he found out that Tartaglia was not the first person to know the method anyway; del Ferro was.

Tartaglia challenged the doctor to a duel. A real duel, not a mathematical one. Cardano never responded, but one of his servants, Ludovico Ferrari, did. Now Ferrari wasn't just any servant. He had come to Cardano as a 14 year old boy seeking employment. Cardano soon singled him out and taught the youth all the mathematics he knew. Ferrari publicized the fact that Tartaglia was not the first person to solve cubic equations, thus ending the threat of a real duel.

Why didn't Tartaglia issue a mathematical challenge to Ferrari?

The answer is simple. Ferrari not only knew the formula for cubic equations, he also knew how to solve fourth degree equations. Tartaglia would certainly lose any duel with Ferrari.

Cardano's book contains the proper acknowledgements to Tartaglia for the formula for cubic equations and to Ferrari for fourth degree equations. *The Great Art* was destined to be the greatest work in algebra for 280 years.

On the cover of this book Tartaglia and a challenger, armed only with an abacus (the calculator of the ancient world), "duel" with mathematical equations while their students look on. The mathematical duels of the Renaissance were gala occasions.

Historical note by David Zitarelli
Illustration by Jay Flom

Copyright © 1990 by Key Curriculum Project, Inc. All rights reserved.
® *Key to Fractions, Key to Decimals, Key to Percents, Key to Algebra, Key to Geometry, Key to Measurement,* and *Key to Metric Measurement* are registered trademarks of Key Curriculum Press.
Published by Key Curriculum Press, 1150 65th Street, Emeryville, CA 94608
Printed in the United States of America   19 18 17   02   ISBN 1-55953-003-0

Solve each equation.

$6x = 18$ $x = 3$	$4s = 24$	$5x = 30$	$9b = 72$
$4y = 8$	$5x = 25$	$3x = {}^-12$	$5s = 0$
$6t = 6$	$^-4n = {}^-20$	$8c = {}^-8$	$^-10e = {}^-30$
$^-7x = {}^-21$	$^-2x = 10$	$20x = 80$	$^-7m = 7$

Robin multiplied 8 times a number. The answer was 48. What was Robin's number?

Equation:

Solution:

Jerry thought of a number. Then he multiplied it by 5. The answer came out to be 45. What was Jerry's number?

Equation:

Solution:

Jennifer started out with 8 dollars. Then she got some more money for her birthday. She ended up with 15 dollars. How much did she get for her birthday?

Equation:

Solution:

7 times some number is 28. What is the number?

Equation:

Solution:

Solving Equations

The equations on the last two pages are very easy to solve. For example, we just have to look at

$$5x = 30$$

to tell that the solution is 6. However, not all equations are that simple. What if we had to solve this one?

$$5x - 3 = 2x + 9$$

One way we can try to solve this equation is by substituting different numbers for x and seeing if they work. Let's see if we can find a number that works.

Let's try 1:

$5x - 3$	$= 2x + 9$
$5(1) - 3$	$2(1) + 9$
$5 - 3$	$2 + 9$
2	11

These two numbers are not the same, so 1 is not the solution.

Try 2:

$5x - 3$	$= 2x + 9$
$5(2) - 3$	$2(2) + 9$
$10 - 3$	$4 + 9$
7	13

2 doesn't work.

Try 3:

$5x - 3$	$= 2x + 9$
$5(3) - 3$	$2(3) + 9$
$15 - 3$	$6 + 9$
12	15

3 doesn't work.

Try 4:

$5x - 3$	$= 2x + 9$
$5(4) - 3$	$2(4) + 9$
$20 - 3$	$8 + 9$
17	17

Both sides came out the same, so 4 is the solution.

Use a pencil and eraser when you do these problems. Try to find the solution to each equation by substituting different numbers until you find one that works.

$4x + 8 = 20$	
$4(3) + 8$	20
$12 + 8$	
20	

The solution is ___3___.

$6x - 8 = 22$	

The solution is _____.

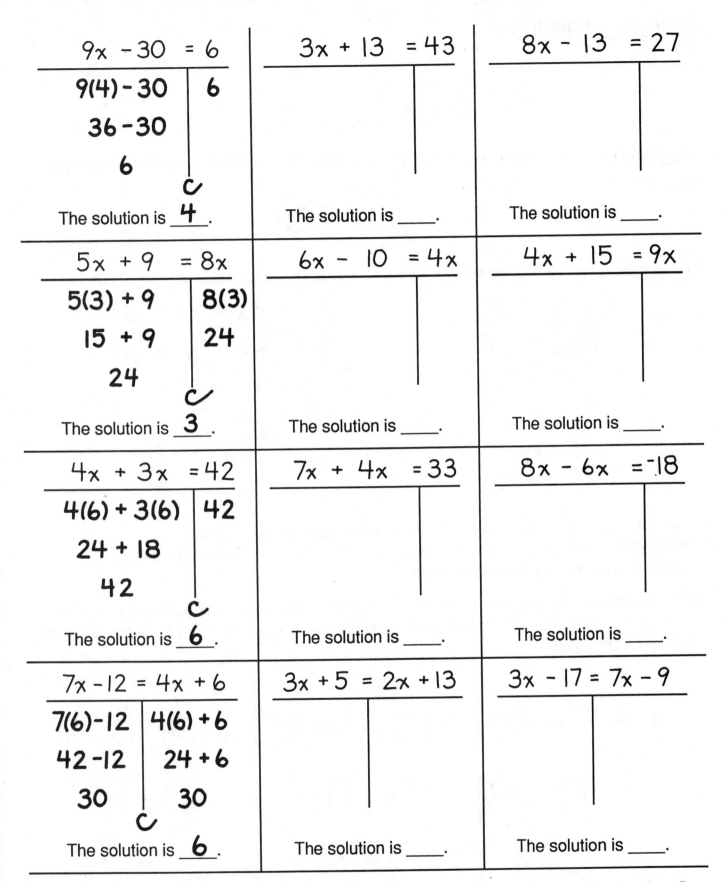

$9x - 30 = 6$		$3x + 13 = 43$	$8x - 13 = 27$
$9(4) - 30$	6		
$36 - 30$			
6			

The solution is __4__. The solution is ____. The solution is ____.

$5x + 9 = 8x$		$6x - 10 = 4x$	$4x + 15 = 9x$
$5(3) + 9$	$8(3)$		
$15 + 9$	24		
24			

The solution is __3__. The solution is ____. The solution is ____.

$4x + 3x = 42$		$7x + 4x = 33$	$8x - 6x = {}^-18$
$4(6) + 3(6)$	42		
$24 + 18$			
42			

The solution is __6__. The solution is ____. The solution is ____.

$7x - 12 = 4x + 6$		$3x + 5 = 2x + 13$	$3x - 17 = 7x - 9$
$7(6) - 12$	$4(6) + 6$		
$42 - 12$	$24 + 6$		
30	30		

The solution is __6__. The solution is ____. The solution is ____.

Did you find solutions to all the equations? Well, don't worry if you didn't. Guessing is a fine way to solve short equations, but it doesn't work well on long ones. The rest of this booklet will show you how to solve many kinds of equations — short ones and long ones, too.

Equivalent Equations

Look at this equation:

$$2x + 3x + 5x = 20$$

We already know that $2x + 3x + 5x$ can be simplified to $10x$. So the equation above can be rewritten as

$$10x = 20$$

Now it is easy to see that the solution to each of these equations is 2.

$10x = 20$
$10(2)$ 20
20

$2x + 3x + 5x = 20$
$2(2) + 3(2) + 5(2)$ 20
$4 + 6 + 10$
20

Whenever two equations have the same solution, we say that they are **equivalent equations**. Since $2x + 3x + 5x = 20$ and $10x = 20$ both have the solution 2, they are equivalent equations.

When we are solving an equation, we do it in **steps**. Each step is an equivalent equation that is easier to solve. In our last step, the equation is so simple that it tells us the solution. Here are some examples:

$2x + 3x + 5x = 20$	$5x - 3x = 4 + 10$	$12 - 36 = 13y - 7y$
$10x = 20$	$2x = 14$	$^-24 = 6y$
$x = 2$	$x = 7$	$y = ^-4$

6

Here are some equations for you to solve. Each problem takes two steps.
First simplify the equation by combining like terms. Then find the solution.

$5x + 2x = {}^-21$ $7x = {}^-21$ $x = {}^-3$	$2x + 3x = 30$	$3x + 3x = {}^-24$	$5x + 5x = {}^-70$
$4x + 5x = 18$	$7x - 3x = {}^-12$	$8x - 14x = 60$	$3x + x = 36$
$5x = 13 + 7$ $5x = 20$ $x = 4$	$3x = 10 + 2$	$^-4x = 19 + 9$	$7x = 30 + 40$
$6x = 5 - 17$	$8x = 30 - 6$	$2x = 23 - 23$	$3x = 11 + 22$
$56 = 2x - 9x$ $56 = {}^-7x$ $x = {}^-8$	$77 = 6t + 5t$	$49 - 9 = 5a$ $40 = 5a$ $a = 8$	$17 + 3 = 4x$
$18 = 7s + 2s$	$15 = 3m - 8m$	$20 - 44 = 8x$	$37 - 30 = 7x$
$20 = 7y - 12y$	$18 = 9t - 3t$	$5 + 13 = {}^-9w$	$48 + 16 = 32y$

Solve each equation.

$5x + 3x = 19 + 5$ $8x = 24$ $x = 3$	$6x + 4x = 15 + 5$	$2x - 7x = 23 + 12$
$10x - 17x = 8 - 50$	$2x + x = 9 + 9$	$5x - 3x + 6x = 56$
$4x + 7x - 4x = 56$	$5x + 2x + 2x = 45$	$6x - 14x + 3x = 40$
$10 + 4 = 2a + 5a$	$3 + 15 = 12b - 3b$	$9 = 10r - 16r + 5r$
$12 + 36 = 13y - 7y$	$^-7 = 4x + 2x + x$	$27 = 6a + 5a - 2a$
$9x - 4x - 8x = 15$	$72 = 4c + 7c - 2c$	$15z - 6z = 20 + 16$
$37 + 35 = 9s + 3s$	$6x - 10x = 28$	$6 = 11y - 4y - 5y$

8

The Finger Method

Look at this equation:

$$3x + 3 = 18$$

It looks very easy to solve, but there's one big problem: $3x$ and 3 are unlike terms, so you can't simplify $3x + 3$. Below is a way to solve equations like this one.

Cover up the $3x$ with your finger:

$$+ 3 = 18$$

What number has to be under your finger to make the equation true? The answer is 15, so

$$3x = 15$$
$$x = 5$$

This is how to write out the solution:

$$+ 3 = 18$$
$$3x = 15$$
$$x = 5$$

Check: $3x + 3 = 18$

$3(5) + 3 \mid 18$

$15 + 3$

18

Here's another equation. You solve this one using the Finger Method.

$$+ 2 = 57$$
$$11x =$$
$$x =$$

Check: $11x + 2 = 57$

Use the Finger Method to solve each equation.

$5x + 2 = 37$ $5x = 35$ $x = 7$	$4x + 7 = 15$	$3x + 8 = 29$
$5x + 6 = 26$	$2x + 10 = 16$	$6x + 7 = 19$
$11 + 4x = 15$ $4x = 4$ $x = 1$	$7 + 3x = 7$	$3 + 8x = 35$
$12 = x + 5$	$4 = 3x + 10$	$87 = 10x + 7$

Here are some more equations to solve using the Finger Method.

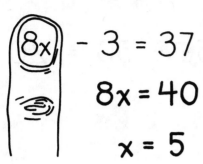

$8x$ $- 3 = 37$

$8x = 40$

$x = 5$

$6x - 11 = 13$

$5x - 14 = 6$

$20 = 3x - 7$

$^-2x + 10 = 24$

$^-2x = 14$

$x =$

$^-7x + 9 = 79$

$10 = 18 -$ $2x$

$2x = 8$

$x = 4$

$6 = 46 - 8x$

$35 = ^-4x + 15$

$15 = 35 - 4x$

You will have to think a little harder to solve these using the Finger Method.

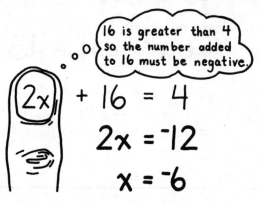

16 is greater than 4 so the number added to 16 must be negative.

$$2x + 16 = 4$$
$$2x = {}^-12$$
$$x = {}^-6$$

$$5x + 50 = 20$$

$$7x + 60 = 4$$

$$19 + 12x = 7$$

$$16 = 24 + 2x$$

$$45 = x + 48$$

$$^-5x + 13 = {}^-12$$

$$^-7x - 1 = {}^-22$$

Did you solve the last two equations? Don't worry if you didn't. The Finger Method is just one way to solve equations. You can keep on using it if you like it. In the next section you will find out about another method which will make it easier to solve equations like the last two.

12

The Addition Principle for Equations

The Addition Principle for Equations says that if we add the same number to each side of an equation, then the new equation we get will be equivalent to the equation we started with. Let's see how this works.

Start with an equation . . .	$6x - 5 = 19$
We can't simplify $6x - 5$, so we need to get rid of the -5. To do this, add 5 to each side of the equation . . .	$6x - 5^{+5} = 19^{+5}$
$-5 + 5 = 0$ and $19 + 5 = 24$, so here's what we have now . . .	$6x = 24$
The rest is easy . . .	$x = 4$

In the example above, we started with the equation $6x - 5 = 19$. Then we added 5 to each side and got the equation $6x = 24$. The Addition Principle for Equations says that these two equations will be equivalent. You can see that they really are equivalent, because 4 is the solution of each.

$$
\begin{array}{c|c}
6x - 5 & = 19 \\
\hline
6(4) - 5 & 19 \\
24 - 5 & \\
19 & \\
& \circlearrowleft
\end{array}
\qquad
\begin{array}{c|c}
6x & = 24 \\
\hline
6(4) & 24 \\
24 & \\
& \circlearrowleft
\end{array}
$$

Here are two more examples:

> We want to get rid of the -7, so add the <u>opposite</u> of -7 to each side of the equation.

$$4x - 7 = 5$$
$$4x - 7^{+7} = 5^{+7}$$
$$4x = 12$$
$$x = 3$$

> We want to get rid of the $+4$, so add the <u>opposite</u> of $+4$ to each side of the equation.

$$8x + 4 = 60$$
$$8x + 4^{-4} = 60^{-4}$$
$$8x = 56$$
$$x = 7$$

Solve each equation by using the Addition Principle.

$$7x - 3 = 11$$

We want to find out what 8x equals, so we need to get rid of the ⁻4. That's why we add ⁺4 to each side.

$$8x - 4^{+4} = 20^{+4}$$
$$8x = 24$$
$$x = 3$$

$$4x - 3 = 21$$

$$5x - 31 = 4$$

$$2x - 9 = 33$$

$$10x - 73 = 7$$

$$11x - 9 = 35$$

$$13^{+23} = 3x - 23^{+23}$$

$$26 = x - 8$$

$$45 = 6x - 9$$

$$6x + 8 = 38$$

We want to find out what 5x equals, so we need to get rid of the ⁺3. That's why we add ⁻3 to each side.

$$5x + 3^{-3} = 23^{-3}$$
$$5x = 20$$
$$x = 4$$

$$10x + 3 = 63$$

$$3x + 7 = 25$$

$$2x + 9^{-9} = 9^{-9}$$
$$2x = 0$$
$$x = 0$$

$$7x + 16 = 16$$

$$34 = 4x + 34$$

14

The problems on this page have been mixed up. Stop and think about what number you want to get rid of. Then add the opposite of that number to both sides.

$4x - 17 = 7$	$6x + 5 = 59$	$5y - 7 = 13$
$2x + 1 = 51$	$3x - 2 = 4$	$10s + 9 = 49$
$9x + 8 = 89$	$5a - 5 = 5$	$6x + 5 = 5$
$32^{+17} = 7m - \cancel{17}^{+17}$ $49 = 7m$ $m = 7$	$50^{-10} = 20x + \cancel{10}^{-10}$ $40 = 20x$ $x = 2$	$35 = 3x + 20$
$22 = 9x - 23$	$3 = 58 + 5x$	$4 = 12x + 40$
$18 = 4z - 10$	$38 = 4x + 10$	$34 = 4x + 10$
$2 = 4x - 10$	$10 = 4x + 10$	$2 = 4x + 10$

Now look at the last two problems on page 12 again. It is easy to solve these using the Addition Principle.

$$-5x + \cancel{13}^{-13} = -12^{-13}$$
$$-5x = -25$$
$$x = 5$$

$$-7x - \cancel{1}^{+1} = -22^{+1}$$
$$-7x = -21$$
$$x = 3$$

Here are some more equations to solve using the Addition Principle:

$-3x + 17 = 2$	$-4x + 6 = -10$	$-5x - 18 = -3$
$-9 + 2x = -21$	$-7x + 42 = 0$	$12 + 9x = -60$
$-70 + 5x = -20$	$-2x - 98 = 2$	$36 = -4x + 56$
$6x + 12 = 0$	$29 = 38 + x$	$0 = 12p - 48$
$18 + -8p = -30$	$0 = -6x - 66$	$-14 + -3x = -2$

16

Sandy and Terry got different answers when they tried to solve the same equation.

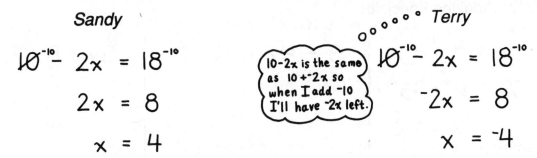

Sandy

$$\cancel{10}^{-10} - 2x = 18^{-10}$$

$$2x = 8$$

$$x = 4$$

Terry

$$\cancel{10}^{-10} - 2x = 18^{-10}$$

$$^-2x = 8$$

$$x = ^-4$$

10-2x is the same as 10 +⁻2x so when I add ⁻10 I'll have ⁻2x left.

Sandy's answer is not a solution. Terry's answer *is* a solution.

$10 - 2x = 18$	
$10 - 2(4)$	18
$10 - 8$	
2	
	X

$10 - 2x = 18$	
$10 - 2(^-4)$	18
$10 - ^-8$	
$10 + 8$	
18	c

Below are some equations for you to solve. Remember to use Terry's method.

$23 - 4x = 27$	$12 - 5x = 27$	$6 - 8x = 30$
$32 - 4x = 0$	$40 - 3x = 34$	$15 - x = 20$
$0 = 18 - 2x$	$^-4 - 7x = 31$	$^-3 - 6x = ^-21$

These equations have an extra step. First simplify each side by combining like terms. Then use the Addition Principle.

$\underline{2x} + \underline{5x} + 11 = 53$ $7x + \cancel{11} = 53$ $7x = 42$ $x = 6$	$3 + 4x + 5x = 21$	$5x + 8 - 2x = 26$
$16x - 8 - 9x = 27$	$x - 3x + 9 = 27$	$2x + 7 + 2x = 23$
$4x + 9x + 8 = 47$	$55 - 3x - 2x = 10$	$8x + 7 + 2x = 7$
$61 = 6 - 8x - 3x$	$x + 6 + x = 28$	$4x - 3 + 2x = 15$
$6x + 14 + x = 56$	$0 = 3x - 20 + 2$	$2x + 5 - 10x + 11 = 0$

On these problems you can use the Addition Principle to get rid of the extra x-terms so that you'll have an x-term on one side and a number term on the other side.

$5x^{-2x} = 2x^{-2x} 12$ $3x = {}^-12$ $x = {}^-4$	$9x = 5x + 8$	$8x = 16 + 6x$	$4x = 7 + 3x$
$10x = 3x - 42$	$7x = 36 + 4x$	$3x = 11x + 24$	$9x = 15 + 6x$
$2x^{+3x} = 45 - 3x^{+3x}$ $5x = 45$ $x = 9$	$4x = 63 - 3x$	$4x = 30 - x$	$2x = 36 - 2x$
$x = 16 - x$	$3x = 48 - 5x$	$10x = 60 - 5x$	$0 = 35 - 7x$

Be careful on these. First look at the equation and decide what term you need to get rid of. Then use the Addition Principle to get rid of it.

$6x + 5 = 35$	$7x = 27 - 2x$	$10t - 3 = 57$	$7x = 3x + 20$
$9x = 2x + 42$	$w + 16 = 7$	$48 - 2x = 4x$	$3k - 10 = 5k$

Be sure to simplify each side before you use the Addition Principle.

$3x + 5x = 6x + 28 - 2x$ $8x^{-4x} = 4x^{-4x} + 28$ $4x = 28$ $x = 7$	$6x + 5x = 3x + 2x - 54$	$7x - 3x = 6x + 14 - 4x$
$3x + x = 4x + 24 - 6x$	$5x + 7x = 3x + 6x - 36$	$14x - 10x = 3x + 16$
$3x - 5x + 8x = 4x + 18$	$2x + 12x = 3x - 5x$	$5x = 3x + 6x + 28$
$26 + 2x + 6 = 8x - 2x$ $32 + 2x^{-2x} = 6x^{-2x}$ $32 = 4x$ $x = 8$	$x + x + 48 = 3x + 5x$	$9x - 6x + 36 = 7x - 2x$
$8x - x - 40 = 10x + x$	$3x + 3x + 3 = 3x$	$6x + 8x - 36 = 3x + 5x$

On some equations you have to use the Addition Principle *twice* to get the x-terms on one side and the numbers on the other.

$$5x^{-2x} + 6 = 2x^{-2x} + 24$$

If I add $-2x$ to both sides I'll have just 24 on the right.

$$3x + 6^{-6} = 24^{-6}$$

Now I can get rid of the 6 by adding -6.

$$3x = 18$$
$$x = 6$$

$$10x + 5 = 6x + 49$$

$7x + 10 = 2x + 25$	$6x + 16 = 3x + 7$	$13x + 41 = 4x + 5$
$6x + 15 = 2x + 7$	$4x + 8 = 3x + 17$	$9x + 40 = 3x + 40$

$$2a^{+3a} - 8 = 12 - 3a^{+3a}$$
$$5a - 8^{+8} = 12^{+8}$$
$$5a = 20$$
$$a = 4$$

$8m - 15 = 7 - 3m$	$4y - 10 = 5 - y$

$4t - 7 = 14 - 3t$	$4m - 7 = 18 - m$	$8y - 9 = 2 - 3y$

Remember to simplify each side before you use the Addition Principle.

$8x + 7 - 3x = 6x + 19 - 4x$
$5x^{-2x} + 7 = 2x^{-2x} + 19$
$3x + 7^{-7} = 19^{-7}$
$3x = 12$
$x = 4$

$5x - 3x + 8 = 4 + 4x - 6$

$9x + 3x - 9x = 6 + 8x - 11$

$4x + 3x - 7 = 60 - 2x - 13$

$6x - 8 + 2x - 5 = 7 - 2x$

$x + 2 + x + x = 6 - x$

$8x + 5x - 7x + 3 - x = 48$

$4 + 3x = 6x - 8 + 3x - 12$

$3x + 3x + 3x + 3 = 3 + 5x$

$7 - 4x + 3 = x - 16 - 3x$

22

Here are some equations with parentheses for you to solve. First simplify each side of the
equation. Then the rest will be easy.

$3(5x - 3x) + 5 = 47$ $3(2x) + 5 = 47$ $6x + 5^{-5} = 47^{-5}$ $6x = 42$ $x = 7$	$2(7x - 3x) + 4 = 28$	$3(2x + 2x) = 35 + 5x$
$16(x - 1) = 12x + 36$	$6x + 2(x + 7) = 46$	$4x + 18 = 7(x + 3)$

$3x + 3(5x - 7x) = 12 - x$	$9 - 4(7x - 8x) = x - 3$
$3(2x - 3) - 9x - 4 = 2x + 12$	$-2x - 18 = 6(1 - 2x) + 4x$

$$4(3x + x) + 7 - 5x = 8 + (-5)(5x - 6x) + 23$$

Using Equations to Solve Problems

Make up an equation for each problem. Then solve the equation to get the answer.

"I'm thinking of a number. If you multiply it by 6 and then add 7, you will get 55. What is my number?"

Equation: $6x + \cancel{7}^{-7} = 55^{-7}$

$$6x = 48$$
$$x = 8$$

Answer: **8**

"I'm thinking of a number. If you multiply it by 4 and then add 13, you will get 37. What is my number?"

Equation:

Answer:

"I'm thinking of a number. If you multiply it by 8 and then add 17, you will get 33. What is my number?"

Equation:

Answer:

"I'm thinking of a number. If you multiply it by 5 and then take away 7, you will get 53. What is my number?"

Equation:

Answer:

"I'm thinking of a number. If you multiply it by 8 and then subtract 13, you will get 43. What is my number?"

Equation:

Answer:

"I'm thinking of a number. If you multiply it by 5 and then subtract 24, you will get the number I am thinking of. What is my number?"

Equation:

Answer:

Do *you* know what number Debra was thinking of? Can you prove it?

24

Here's how we could use algebra to solve the Shake Shack problem from page 1. Remember that a shake at the Shack costs 80¢ and the bill for three burgers and a shake was $4.40. The question was "How much is a burger?"

First we pick a variable to stand for a number we want to find.

x = cost of a burger

Next we use that variable to write expressions for other numbers we will need.

$3x$ = cost of 3 burgers

Then we find two things which are equal and write an equation.

$3x + 80 = 440$ ($4.40 is 440¢)

$$3x = 360$$
$$x = 120$$

Finally we solve the equation and use the solution to answer the question.

A burger costs $1.20.

Follow the steps above to solve each problem.

At the Shake Shack two orders of fries and four shakes costs $5.90. How much is an order of fries? Equation: Answer:	Jed figures the prom will cost him $160. He has saved $46 and can earn $6 an hour at his job. How many hours will he have to work? Equation: Answer:
Matt wants to call his mother long distance. The rate is 80¢ for the first three minutes and 20¢ for each additional minute. How long can he talk for $3.00? Equation: Answer:	Anna is in a phone booth with $1.35 in change. A call home costs 60¢ for the first five minutes and 15¢ for each additional minute. How long can she talk? Equation: Answer:

For each problem, first write an equation. Then solve the equation and use the solution to find the answer.

A 48-cm piece of wire is to be cut into two parts. One piece must be 10 cm longer than the other. How long should the pieces be?	How can $500 be divided between two people so that one person gets $50 more than the other?

$$\overset{\displaystyle x \quad , \quad x + 10}{\underline{}}$$
$$48$$

Equation: $x + x + 10 = 48$

$$2x + 10 = 48$$
$$2x = 38$$
$$x = 19$$

Answer: 19 cm and 29 cm

Equation:

Answer:

A 16-ft. board is to be cut into three pieces. Two of the pieces must be 4 feet shorter than the third piece. How long should each piece be?	The band leader wants to line up the 92 band members in 4 rows so that each row has two more members than the row before. How many members should be in the shortest row?

Equation:

Answer:

Equation:

Answer:

Raoul has to finish a 473-page book in a week. He decides to read the same number of pages each weekday and 30 extra pages on Saturday and on Sunday. How many pages will he have to read on Wednesday?	A store owner sells two pairs of running shoes for each pair of high tops. He plans to order 144 pairs of shoes How many pairs should be high tops?

Equation:

Answer:

Equation:

Answer:

Age Puzzles

Algebra makes it easy to solve certain kinds of puzzles. Here are some about age.
Follow the same steps you followed on the last two pages.

Jason is 12 years older than Ted. Next year Jason will be 3 times as old as Ted. How old is Ted?

	Ted	Jason
This year:	x	$x+12$
Next year:	$x+1$	$x+13$

Equation: $x+13 = 3(x+1)$

$$x+13 = 3x+3$$
$$13 = 2x+3$$
$$10 = 2x$$
$$x = 5$$

Answer: Ted is 5 years old.

Ellen is 11 years older than Maja.
Last year Ellen was twice as old as Maja.
How old is Maja now?

Equation:

Answer:

Pam is 14 and her dad is 37. In how many years will Pam's dad be twice as old as she will be?

Equation:

Answer:

Sean is 20 and his brother is 12.
How many years ago was Sean three times as old as his brother?

Equation:

Answer:

Minh is 16. His parents are both the same age. The three of them have lived a total of 100 years. How old are his parents?

Equation:

Answer:

Alex and Alicia are twins. Kevin is 5 years older than the twins. Their ages total 53. How old are the twins?

Equation:

Answer:

Perimeter Problems

In each problem, first write an equation for the perimeter of the rectangle or triangle. Then solve the equation and use your solution to find the lengths of all the sides.

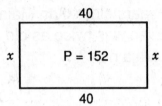

Equation: $x + 40 + x + 40 = 152$

$2x + 80 = 152$

$2x = 72$

$x = 36$

Lengths: 36, 40, 36, 40

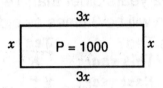

Equation:

Lengths:

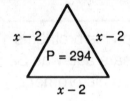

Equation:

Lengths:

25 | P = 143 | x − 6
x − 6

Equation:

Lengths:

You need to make your own sketches for these problems.

The perimeter of a rectangle is 234 meters. The rectangle is twice as long as it is wide. What are its length and width?

Equation:

Answer:

Zelda has a 300-inch roll of binding for a quilt. The quilt must be 84 inches long. How wide can she make it?

Equation:

Answer:

The Division Principle for Equations

In arithmetic there are three ways to write the division problem "32 divided by 4":

$$32 \div 4 \qquad 4\overline{)32} \qquad \frac{32}{4}$$

In algebra we usually write it the third way: $\frac{32}{4}$

Find an answer to each division problem below. If you can't do the problem in your head, use a calculator or do the arithmetic on a piece of scratch paper.

$$\frac{72}{3} = \qquad\qquad \frac{^-7360}{^-10} = \qquad\qquad \frac{187}{^-11} =$$

$$\frac{^-96}{4} = \qquad\qquad \frac{600}{25} = \qquad\qquad \frac{742}{53} =$$

$$\frac{^-57}{^-3} = \qquad\qquad \frac{^-1024}{32} = \qquad\qquad \frac{^-976}{^-122} =$$

$$\frac{^-510}{6} = \qquad\qquad \frac{4080}{^-30} = \qquad\qquad \frac{0}{329} =$$

The bar we use to show division is like parentheses. Do whatever is above or below the bar *before* doing the division.

$$\frac{5 + 9}{3 + 4} = \qquad\qquad \frac{1 - 37}{2 - 20} = \qquad\qquad \frac{6 + 3}{6 - 3} =$$

$$\frac{7(^-4)}{2} = \qquad\qquad \frac{^-6 \cdot 8}{16} = \qquad\qquad \frac{4 \cdot 9}{6 \cdot 6} =$$

$$\frac{3 \cdot 7 - 1}{^-2 \cdot 5} = \qquad\qquad \frac{(^-9)^2}{3^2} = \qquad\qquad \frac{3^2 + 4^2}{1^2 + 2^2} =$$

$$\frac{4 \cdot 9}{6 + 6} = \qquad\qquad \frac{28 \cdot 1000}{30 + 5} = \qquad\qquad \frac{300 - 5}{10 - 5} =$$

Some equations, like $5x = 30$, are easy to solve in your head. When the numbers in the equation are larger we can use the Division Principle to find the solution.

The Division Principle says that if we divide both sides of an equation by the same number, the new equation we get will be equivalent to the equation we started with.

$$3x = 192$$

$$\frac{\cancel{3}x}{\cancel{3}} = \frac{192}{3} \longrightarrow \quad 3\overline{)192}$$

$$x = 64$$

$$\begin{array}{r} 64 \\ 3\overline{)192} \\ \underline{18} \\ 12 \\ \underline{12} \\ 0 \end{array}$$

Solve each equation using the Division Principle.

If I multiply by 3, then divide by 3, I get the number I started with: X. $\dfrac{\cancel{3}x}{\cancel{3}} = \dfrac{72}{3}$ $x = 24$	$4x = {}^-96$	$^-3x = {}^-57$
$4x = 112$	$^-12x = {}^-288$	$6x = {}^-654$
$^-455 = 5x$	$536 = 8x$	$^-715 = {}^-11x$
$25x = 600$	$84x = {}^-1260$	$21x = 13419$
$70x = 21000$	$^-1x = 45$	$^-x = 23$

30

Use the Addition and Division Principles to solve each equation.

$5x - 120^{+120} = 60^{+120}$ $\dfrac{5x}{5} = \dfrac{180}{5}$ $x = 36$	$3x - 18 = 45$	$6x + 42 = 300$
$10 - 4x = 102$	$100 - 9x = 451$	$5x - 17 = 123$
$14x + 25 = 193$	$32x - 77 = 115$	$18x + 25 = {}^-65$
$\dfrac{3(x - 2)}{3} = \dfrac{99}{3}$ $x - 2 = 33$ $x = 35$ OR $\begin{array}{l} 3(x - 2) = 99 \\ 3x - 6^{+6} = 99^{+6} \\ \dfrac{3x}{3} = \dfrac{105}{3} \\ x = 35 \end{array}$		$7(x - 5) = 252$
$5(x + 6) = 400$	$-12(x + 4) = 240$	$6(2x + 1) = {}^-138$
$196 + 14x = 0$	$2x - 55 = 11x - 280$	$50 + 4(x - 2) = 210$

The Multiplication Principle for Equations

In some equations the variable is already divided by a number.

$$\frac{x}{5} = 14 \qquad\qquad \frac{x}{25} = {}^-18$$

We can use the Multiplication Principle to solve these equations. The Multiplication Principle says that if you multiply both sides of an equation by the same number (except 0), the new equation you get will be equivalent to the equation you started with.

$$\frac{x}{5} = 14 \qquad\qquad \frac{x}{25} = {}^-18$$

$$5 \cdot \frac{x}{5} = 14 \cdot 5 \qquad 25 \cdot \frac{x}{25} = {}^-18 \cdot 25$$

$$x = 70 \qquad\qquad x = {}^-450$$

$$\begin{array}{r} 18 \\ \times\ 25 \\ \hline 90 \\ 36 \\ \hline 450 \end{array}$$

Solve each equation using the Multiplication Principle.

$6 \cdot \frac{x}{6} = 5 \cdot 6$ $x = 30$	$\frac{x}{-2} = 12$	$\frac{x}{4} = 10$	$\frac{x}{9} = {}^-8$
$\frac{x}{10} = 42$	$\frac{x}{-2} = {}^-8$	$\frac{x}{8} = {}^-2$	$\frac{x}{6} = 8$
$\frac{x}{11} = 2$	$\frac{x}{27} = {}^-1$	$\frac{x}{-10} = {}^-10$	$\frac{x}{15} = 0$
$9 = \frac{x}{9}$	$1 = \frac{x}{24}$	$13 = \frac{x}{-3}$	$11 = \frac{x}{80}$
$\frac{x}{22} = 15$	$\frac{x}{-16} = 18$	$\frac{x}{100} = 34$	$\frac{x}{24} = {}^-56$

Use the Multiplication and Addition Principles to solve each equation.

$\frac{x}{3} - 7^{+7} = 15^{+7}$ $3 \cdot \frac{x}{3} = 22 \cdot 3$ $x = 66$	$\frac{y}{5} - 11 = 9$	$\frac{n}{9} + 5 = 13$
$\frac{r}{-2} + 8 = 5$	$\frac{a}{6} - 15 = 15$	$\frac{d}{10} + 15 = 15$
$12 = 5 + \frac{x}{3}$	$14 = \frac{k}{8} + 7$	$12 + \frac{a}{32} = 22$
$4 \cdot \frac{x-9}{4} = 20 \cdot 4$ $x - 9^{+9} = 80^{+9}$ $x = 89$	$\frac{x-7}{8} = {}^-3$	$\frac{x+5}{3} = 9$
$\frac{x}{5} + 11 = 731$	$\frac{x-8}{33} = 40$	$12 = 18 + \frac{x}{6}$

Solve each equation by using the Multiplication and Division Principles.

$2 \cdot \dfrac{5x}{2} = 30 \cdot 2$ $\dfrac{5x}{5} = \dfrac{60}{5}$ $x = 12$	$\dfrac{10y}{9} = 100$	$\dfrac{7x}{8} = 14$
$\dfrac{2x}{3} = 120$	$\dfrac{-6y}{5} = 18$	$\dfrac{-x}{14} = 2$
$\dfrac{15x}{2} = {}^-75$	$\dfrac{-3x}{4} = 24$	$\dfrac{11x}{12} = 55$
$\dfrac{-50x}{3} = 1000$	$\dfrac{5x}{9} = {}^-40$	$\dfrac{4x}{-1} = 64$
$12 = \dfrac{6x}{5}$	${}^-20 = \dfrac{5x}{6}$	$0 = \dfrac{-2x}{3}$
$\dfrac{3(x-5)}{2} = 9$	$\dfrac{2(x+8)}{3} = 16$	$\dfrac{3(2x-1)}{7} = 9$

34

Written Work

Do these problems on some clean paper. Label each page of your work with your name, your class, the date, and the book number. Also number each problem. Keep this written work inside your book, and turn it in with your book when you are finished. Please do a neat job.

Solve each equation.

a)
$3x = 15$ $4 = 2x$ b) $3b + 2b = 10$ $2p + 4p = 6$

$3a = 9$ $^-7a = ^-21$ $35 = 5x + 2x$ $^-5 + 9 = 2y$

$13y = ^-26$ $7y = 0$ $2a - 3a = 4$ $2a - 3a = ^-4$

$^-12 = 2c$ $6n = 186$ $9x = 12 + 627$ $7x - x = 30$

$^-25 = 5x$ $14y = ^-126$ $7x - 2x = ^-2005$ $5x - 8x = ^-15$

c)
$2x - 2 = 6$ $7x = 14 + 5x$ $7 = b + 6$ $5 = 6p - 7$

$4a - 3 = 9$ $2x = ^-4x + 6$ $x + 7 = 3$ $8 = 3y + 2$

$13 = 7x - 1$ $2x = ^-3 + x$ $0 = 3t + 21$ $^-6x = 3x$

$7 = 2p + 5$ $a - 6 = 3a$ $7r = 5r$ $^-5 = ^-2 - 3x$

$2t - 8 = 0$ $3x = 8 - x$ $^-20 = 1 - 7t$ $6 - 2x = ^-14$

d)
$2x + 5 = 5x - 4$ e) $3y + 2 - 4y = 6$ $0 = 5x + 7 - 2x - 16$

$^-3x + 4 = 2x + 24$ $3 = 6x - 3 - 3x$ $6a - 4 + 2 = 3a + 1$

$3x - 4 = 4x + 2$ $30 = 6r - 24 + 3r$ $2y - 3 + 3y = 4y + 2$

$7 - 2y = 1 + y$ $x + 13 - 8x = ^-1$ $5x + 3 = 2x - 3x - 15$

$6 - 3x = 2 + x$ $4x + 7 + 2x = ^-5$ $3x + 4 - 5x + 2 = 0$

$6x - 5 = 2x - 5$ $^-2x - 5x - 4 = 2x + 12 - x$ $10x - 9 - 9 = 14x - 9 + 5x$

f) $3(2x + x) - 11 = 7$ g) $\dfrac{x}{8} - 3 = 10$ $\dfrac{3x + 2}{5} = 10$ $5(x + 4) = 95$

$0 = 4(2y - 4y) + 2y - 24$

$(7y - 8y)2 + 14 = ^-2$ $\dfrac{x + 4}{6} = ^-20$ $\dfrac{x}{2} - 23 = 41$ $\dfrac{x - 16}{45} = 28$

$7(2x + 3) = 35$

$5(x + 6) + 24 = 9$ $\dfrac{x - 1}{12} = 7$ $11 + \dfrac{x}{3} = 28$ $\dfrac{8(x - 1)}{4} = 36$

$x + 3(2x - 4) = ^-19$

Practice Test

Solve each equation.

$x + 10 = 14$	$x + 7 = 13$	$^-5x = 30$	$8x = {}^-56$
$x + 4 = 4$	$x + 9 = 0$	$10x = 10$	$12x = 0$

$2x - 3 = 19$	$8x + 15 = {}^-1$	$5 - 4x = 45$
$7x = 2x + 30$	$12 - x = 3x$	$9x - 10x = 14$

$3(x + 3) + 4x = 15 - 2x + 3$	$^-6x + 2 - 5x + 7 + x = 39$
$x - 5 + x - 6 = 3x - 7$	$5x + 2 - 6x - 10 = x + 520$

Solve each equation.

$16x = 560$	$^-22x = 440$	$^-7x = ^-448$
$\frac{x}{12} = 5$	$\frac{x}{3} - 1 = 20$	$\frac{x}{^-5} = 100$
$\frac{x+6}{6} = 14$	$\frac{x-10}{25} = ^-2$	$7(x-3) = 35$

Make up an equation for each problem. Then solve the equation to get the answer.

"I'm thinking of a number. If you multiply it by 6 and then add 8, you will get 32. What is my number?"

Equation:

Answer:

A 98-inch piece of wire must be cut into two pieces. One piece must be 10 inches shorter than the other. How long should the pieces be?

Equation:

Answer:

First class postage costs 25¢ for the first ounce and 20¢ for each additional ounce. How many additional ounces are you being charged for if the postage is $1.05?

Equation:

Answer:

Find the lengths of the sides of this rectangle.

$x + 9$

x P = 302 x

$x + 9$

Equation:

Answer:

Key to Algebra®

Book 1: *Operations on Integers*
Book 2: *Variables, Terms and Expressions*
Book 3: *Equations*
Book 4: *Polynomials*
Book 5: *Rational Numbers*
Book 6: *Multiplying and Dividing Rational Expressions*
Book 7: *Adding and Subtracting Rational Expressions*
Book 8: *Graphs*
Book 9: *Systems of Equations*
Book 10: *Square Roots and Quadratic Equations*
Answers and Notes for Books 1–4
Answers and Notes for Books 5–7
Answers and Notes for Books 8–10

Also Available

Key to Fractions®
Key to Decimals®
Key to Percents®
Key to Geometry®
Key to Measurement®
Key to Metric Measurement®

KEY CURRICULUM PRESS
Innovators in Mathematics Education

100% PCW

ISBN 1-55953-003-0

90000>

9 781559 530033

TUBE FLIES TWO
Evolution

Mark Mandell and
Bob Kenly

Color Plates by Craig Wester

Frank
Amato
PORTLAND

DEDICATION

This book is for the boys:
Sam and Nat Mandell, and Matt Carter.

Mark Mandell

Fly-tying requires concentration,
and often physical and mental isolation.
My wife Betty, who is an artist, understands the need
for solitude and forced absences. I dedicate this work
to her with all the love I have to give.

Bob Kenly

◆ ◆ ◆

ACKNOWLEDGMENTS

The authors are indebted to Sandy Leventon, editorial consultant
of *Trout and Salmon* magazine (UK), John Albright of HMH/Kennebec River
Fly and Tackle, Martin Joergensen of GlobalFlyfisher.com, Rex Andersen,
and Arthur Greenwood. Without their help in identifying and
locating far-flung tube tiers, this compendium would have
been little more than a pamphlet.

Photographs by the author unless otherwise noted.

Color plates by Craig Wester

Book & Cover Design: Kathy Johnson

Printed in Singapore

Softbound ISBN: (10) 1-57188-401-7 (13) 978-1-57188-401-5 UPC: 0-81127-00235-1

Hardbound ISBN: (10) 1-57188-402-5 (13) 978-1-57188-402-2 UPC: 0-81127-00236-8

1 3 5 7 9 10 8 6 4 2